spot

OUR SOLAR SYSTEM

# SATURN

by Alissa Thielges

AMICUS

rings

hexagon

Look for these words and pictures as you read.

probe

Titan

Look at that giant planet.
It is Saturn!

Saturn is very cold.

It is the sixth planet from the sun.

Neptune
Uranus
Saturn
Jupiter

See the rings? There are seven. They are made of rocks and ice.

rings

See the hexagon storm?
It has six sides.
It is at the top of Saturn.

**hexagon**

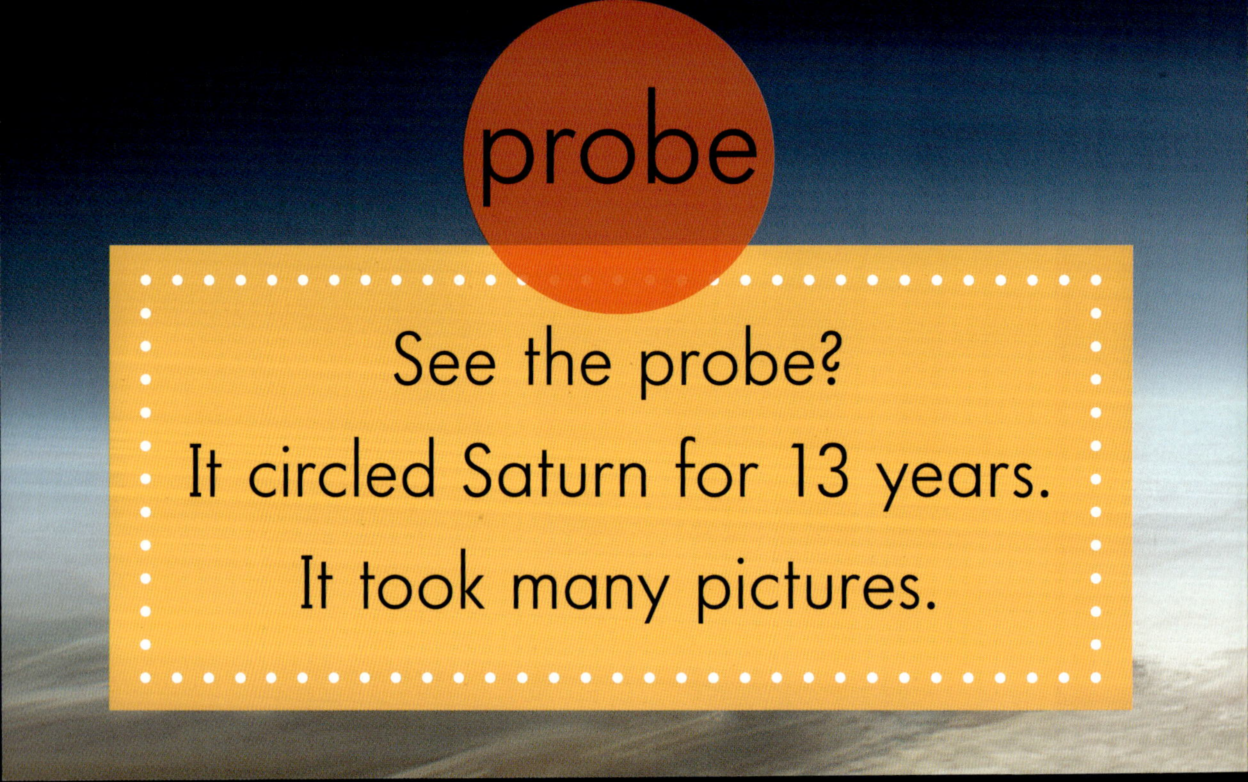

## probe

See the probe?
It circled Saturn for 13 years.
It took many pictures.

See the moon, Titan?
Saturn has over 80 moons.
Titan is the biggest.

Look! You can see Saturn from Earth! It looks like a star.

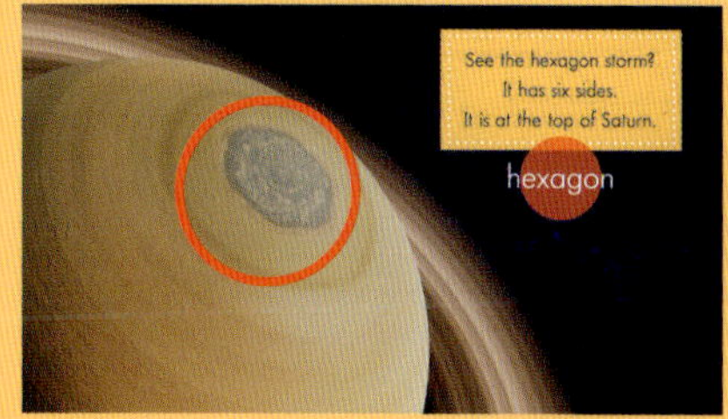


rings

hexagon

# Did you find?

probe

Titan

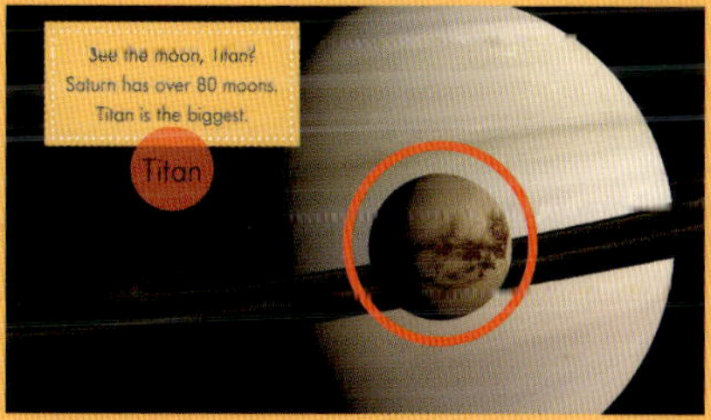

Spot is published by Amicus Learning, an imprint of Amicus
P.O. Box 227, Mankato, MN 56002
www.amicuspublishing.us



Library of Congress Cataloging-in-Publication Data
Names: Thielges, Alissa, 1995– author.
Title: Saturn / by Alissa Thielges.
Other titles: Spot. Our Solar System.
Description: Mankato, MN : Amicus, [2024] | Series: Spot. Our Solar System | Audience: Ages 4–7 | Audience: Grades K–1 | Summary: "Simple text and a search-and-find feature reinforce new science vocabulary about Saturn's weather, moons, and latest NASA probe for early readers"—Provided by publisher.
Identifiers: LCCN 2022035873 (print) | LCCN 2022035874 (ebook) | ISBN 9781645492719 (library binding) | ISBN 9781681527956 (paperback) | ISBN 9781645493594 (ebook)
Subjects: LCSH: Saturn (Planet)--Juvenile literature.
Classification: LCC QB671 .T54 2024 (print) | LCC QB671 (ebook) | DDC 523.46—dc23/eng20230106
LC record available at https://lccn.loc.gov/2022035873
LC ebook record available at https://lccn.loc.gov/2022035874

Printed in China

Rebecca Glaser, editor
Deb Miner, series designer
Lori Bye, book designer
Omay Ayres, photo researcher

Photos by Dreamstime/Luyag2 6–7; Getty/ewg3D 4–5, Gary Hershorn 14, SCIEPRO/SCIENCE PHOTO LIBRARY cover; iStock/Ianm35 12–13; NASA/JPL-Caltech 10–11; Shutterstock/Dotted Yeti 3, 16, Pike-28 8–9; Wikimedia Commons/NASA/JPL/Space Science Institute 1

SATURN